THE ELECTROMAGNETIC SPECTRUM

UNDERSTANDING AND USING GAMMA RAYS

ALISON AND STEPHEN ELDRIDGE

Enslow PUBLISHING

Published in 2022 by Enslow Publishing, LLC.
101 W. 23rd Street, Suite 240, New York, NY 10011

Library of Congress Cataloging-in-Publication Data

Names: Eldridge, Alison, author. | Eldridge, Stephen, author.
Title: Understanding and using gamma rays / Alison and Stephen Eldridge.
Description: New York, NY : Enslow Publishing, 2022 | Series: The electromagnetic spectrum | Includes bibliographical references and index. Audience: Grades 7–12.
Identifiers: LCCN 2019008579 | ISBN 9781978514898 (library bound) | ISBN 9781978514881 (pbk.)
Subjects: LCSH: Gamma rays—Juvenile literature.
Classification: LCC QC793.5.G32 E43 2022 | DDC 539.7/222—dc23
LC record available at https://lccn.loc.gov/2019008579

Printed in the United States of America

To Our Readers: We have done our best to make sure all websites in this book were active and appropriate when we went to press. However, the author and the publisher have no control over and assume no liability for the material available on those websites or on any websites they may link to. Any comments or suggestions can be sent by email to customerservice@enslow.com.

Photos Credits: Cover Stocktrek Images/Getty Images; pp. 6, 15 VectorMine /Shutterstock.com; p. 8 NASA/Goddard Space Flight Center Conceptual Image Lab; p. 11 CWA Studios/Shutterstock.com; p. 12 udaix/Shutterstock.com; p. 17 MPanchenko/Shutterstock.com; p. 19 Fouad A. Saad/Shutterstock.com; p. 22 Bettmann/Getty Images; p. 25 Qwerty123uiop/Wikimedia Commons/File: PhotoMultiplierTubeAndScintillator.svg/CC BY-SA 3.0; p. 29 X-ray: NASA/CXC /SAO/PSU/D. Burrows et al.; Optical: NASA/STScI; Millimeter: NRAO/AUI/NSF; p. 31 Lionel Flusin/Gamma-Rapho/Getty Images; p. 34 David Parker/Science Source; p. 37 NASA; p. 40 Joe Raedle/Hulton Archive/Getty Images; p. 42 NASA/JPL /University of Arizona; p. 44 The Washington Post/Getty Images; p. 48 baranozdemir/iStock/Getty Images; p. 50 BSIP/Universal Images Group/Getty Images; cover background, interior pages (faceted color pattern) s_maria/Shutterstock.com; interior pages (wave graphic) MagicMary/Shutterstock.com.

CONTENTS

INTRODUCTION

Sunlight shines through the wet air after a thunderstorm, making a beautiful rainbow. An X-ray machine blasts a human body and records an outline of the bones within. A radio antenna picks up a signal from miles away and turns it into music that blares through a car as it flies down the highway. These are all different forms of the same basic concept—electromagnetic radiation. Electromagnetic radiation is a kind of energy created by electromagnetic fields, or fields created by the motion of objects that have a charge, such as electrons. Electromagnetic radiation flows through the universe as waves. It often interacts with other objects as it flows, which is why it can be used by humans for things like communication, medicine, and even eyesight. The many different forms this radiation takes can be described by a simple property called wavelength.

Wavelength, Frequency, and Energy

Imagine a very still pond or pool. Its surface is very smooth and flat. What happens when a stone is dropped into it? Small waves ripple through the water, creating little peaks and valleys in it. These kinds of waves, known as mechanical waves, work by bumping molecules into each other like dominoes. Any given molecule of water moves very little, but a wave itself can travel across the pond very quickly.

Suppose you want to describe these waves to someone who couldn't see them. One way to describe waves is to talk about the distance between the peaks created as the wave travels. That distance is called a wavelength.

All waves of electromagnetic radiation are similar. Instead of moving through molecules, the waves move through electromagnetic fields. Because they move through electromagnetic fields, they move at the speed of light. In fact, light itself is a form of electromagnetic radiation. Yet these waves still have peaks and valleys, and people can still measure the length between them to describe a wave.

A related way to describe waves is by their frequency—how many peaks appear in a given time. Smaller wavelengths lead to higher frequencies because more peaks will fit into a given amount of time when the peaks are closer together.

Electromagnetic waves are thought to be carried by a kind of particle called a photon.[1] A photon is a tiny packet of energy. The amount of energy carried by a photon is related to the wavelength of the electromagnetic wave the photon carries. The shorter the wavelength of an electromagnetic wave, the higher the energy carried by its photon.[2]

The wavelength, frequency, and energy of electromagnetic radiation are what separate the light we can see from the energy that cooks food in a microwave and the deadly radiation used to sterilize surgical equipment.[3] The infinite array of different wavelengths of electromagnetic radiation is called the electromagnetic spectrum. The spectrum is broken into different regions. Because wavelength, frequency, and energy are so closely related, scientists classify electromagnetic waves using any of those measurements.

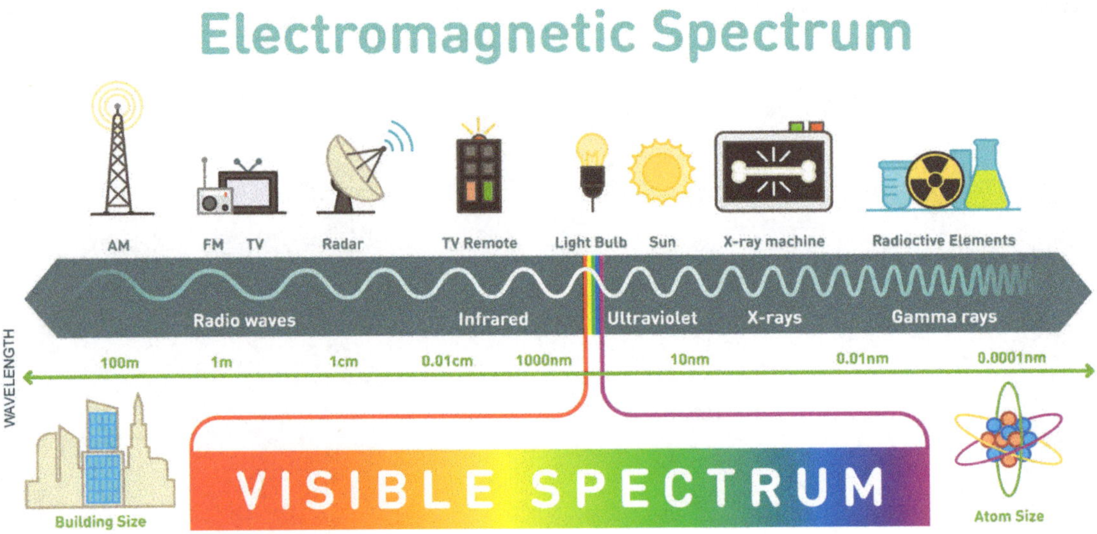

Gamma rays have the shortest wavelengths in the electromagnetic spectrum. The visible spectrum is just a small part of the electromagnetic spectrum.

Gamma rays are one kind of electromagnetic radiation. They are the electromagnetic waves with the shortest wavelengths. Gamma rays usually have a wavelength of a ten-billionth of a meter or less. This means that gamma rays have wavelengths smaller than the diameter of an atom—sometimes much smaller.[4]

Perhaps you are familiar with gamma rays from comic books and superhero movies. However, you may not have known that gamma-ray bursts are the most spectacular explosions observed in the universe. What exactly are gamma rays? How do people encounter them and use them in their daily lives?

WHAT ARE GAMMA RAYS?

CHAPTER ONE

Small wavelengths mean high frequencies and big energy. Gamma rays are the most powerful form of electromagnetic radiation.[1] They move from peak to peak at a frequency of at least ten quintillion cycles per second. That's ten million million million times in one second! There's no known limit to how much energy a gamma ray can carry.

Gamma rays are very similar to X-rays, and the difference between them can be hard to define. In the past, it was all about wavelength. Electromagnetic rays with the shortest wavelengths were called gamma rays. Rays with slightly longer wavelengths were called X-rays. More recently, scientists have begun to distinguish between gamma rays and X-rays based on how they're generated. Gamma rays are generated by the nucleus of an atom, and X-rays aren't. Either way, gamma rays generally have shorter wavelengths and higher energy.

Having high energy gives gamma rays some interesting properties. Most important, it makes them

This conceptual image from NASA shows gamma ray flares that may have occurred when a magnetar's surface cracked. A magnetar is a type of neutron star that has a very strong magnetic field.

very dangerous to humans and other living things. It also means they tend to travel right through solid objects, which makes them useful in medicine in a similar way to X-rays. And like most electromagnetic radiation, you can't see them. Finally, their high power means they're often released by some of the most energetic events in the universe. They travel huge distances across space to be detected by instruments here on Earth.

THE INCREDIBLE HULK: FACT AND FICTION

Gamma rays may be most famous as the power source for the Marvel superhero the Incredible Hulk. Bruce Banner is a scientist caught in the blast of a gamma bomb, which gives him incredible strength.

Unfortunately, the actual effects of gamma radiation are deadly.[2] But gamma rays were an appropriate choice for Marvel's strongest hero. In the comics, there is no limit to the Hulk's strength. There is also no theoretical limit to how strong gamma rays can be.[3] Just like the Hulk, gamma rays are "the strongest one there is."

What Is a Gamma Anyway?

So why are they called "gamma" rays? What does "gamma" mean?

Gamma is actually the third letter of the Greek alphabet. The first two letters are alpha and beta (together, they're the origin of the word "alphabet"). "Gamma" is used to distinguish gamma rays from other types of radiation—specifically, alpha particles and beta particles. Alpha particles, beta particles, and gamma rays are all kinds of ionizing radiation.

Ionizing Radiation

Some radiation can be deadly. Radiation suits are a common feature in movies and on television, and lead-lined aprons are familiar in doctors' and dentists' offices to protect patients and workers from X-rays. But people don't usually protect themselves this way from radio waves or visible light. Clearly, not all radiation is made equal. One of the main things that determines how harmful radiation can be is whether or not it is ionizing radiation.[4]

Human bodies are made up of atoms. The parts of an atom are called protons, neutrons, and electrons. The center of an atom, called the nucleus, is made of some combination of protons and neutrons. The nucleus is surrounded by electrons. These electrons can help bond atoms together into units called molecules.

The molecules that make up the human body generally have no charge. The positive charge of the protons in a molecule cancels out the negative charge of the electrons. This means the total charge is zero. But when an atom or molecule does have a charge—when it doesn't have just the right number of electrons to even out the protons—it is called an ion. Ions can have different properties than neutral molecules.[5] Those electrons make a difference!

Ionizing radiation is radiation that is powerful enough to dislodge electrons from atoms and molecules.[6] When ionizing radiation strikes a human body, it can turn molecules into ions or break up bonds between molecules. In low doses, this can cause nausea, vomiting, and hair loss.[7] In higher doses, it can kill. Even if a person survives an exposure to ionizing radiation, it can have long-term effects on their DNA.[8] These effects make it more likely the person will develop cancer or other harmful mutations. This is why people who work with radiation take precautions to protect themselves from its effects.

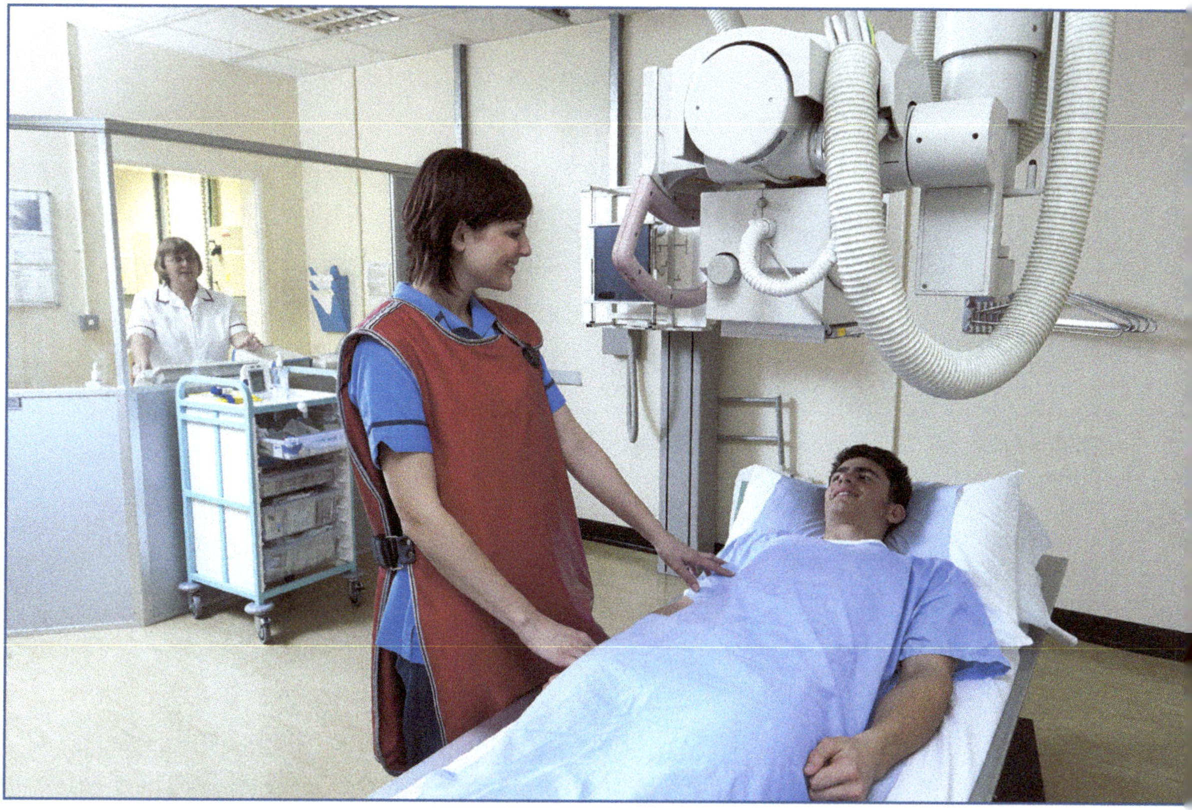

Radiation can be harmful to humans. This technician wears a lead apron when she administers X-rays because she is exposed to a lot of X-ray radiation every day.

Alpha Particles and Beta Particles

High-energy electromagnetic radiation (such as gamma rays, X-rays, and some ultraviolet radiation) is ionizing radiation. However, lower-energy electromagnetic radiation (such as radio waves and visible light) is not ionizing.

Other kinds of ionizing radiation include alpha particles and beta particles. Alpha particles, beta particles, and gamma rays can all be released by a process called radioactive decay.[9]

Radioactive decay is the process by which radioactive materials release radiation over time.

Alpha particles are relatively large and slow. But slow for nuclear radiation means moving at over 20,000 miles per hour (32,187 kilometers per hour)! Alpha particles are made up of two protons and two neutrons bound tightly together. They are called alpha particles because they were the first kind of nuclear radiation discovered.[10]

Beta particles were found next. They are high-energy electrons and move at the speed of light. Gamma rays were found third, so they were named after the third letter of the Greek alphabet. There is one more common kind of ionizing nuclear radiation, but it doesn't have a Greek letter for a name. It's just known as neutron radiation, or neutron emission.[11]

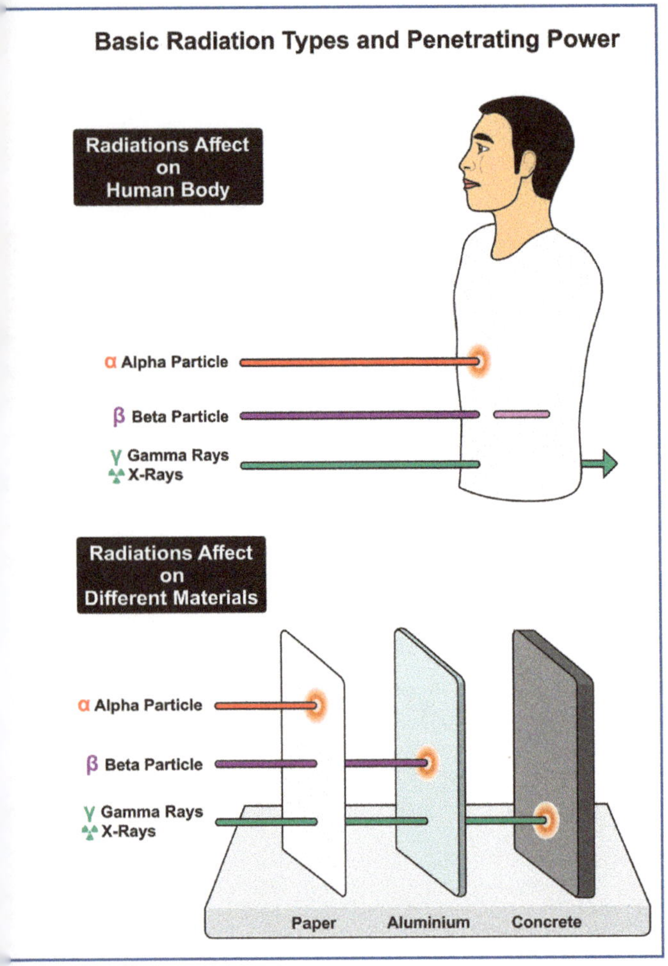

Gamma rays can penetrate the human body, paper, and aluminum. They can be stopped by about a foot of lead, several feet of concrete, or a large quantity of water.

CHAPTER TWO

THE SECRET ORIGINS OF GAMMA RAYS

So, we know that gamma rays are a kind of electromagnetic radiation and, in fact, the most powerful kind. But where do they come from? How is it that all that energy comes into being?

Gamma rays are generally released by nuclear reactions. They can also be produced by interactions between matter and antimatter. Gamma rays are one of the kinds of radiation released by explosive changes to atoms, like the reactions in nuclear power, atomic bombs, and the sun and the stars. Gamma rays are also released by radioactive decay over time.[1]

Radioactive Decay

There are three main kinds of radioactive decay: alpha decay, beta decay, and gamma decay. Alpha and beta decay release alpha and beta particles respectively.

But gamma decay releases gamma rays. Gamma decay often occurs after alpha decay and beta decay. That means that many kinds of radioactive decay will eventually result in the release of gamma rays. Before learning about alpha, beta, and gamma decay, it's important to understand more about the parts of the atom and the roles they play in nuclear reactions.

Nuclear reactions are called "nuclear" because they take place in the nucleus of an atom. Each atom has a nucleus made up of protons and neutrons. Because of this, protons and neutrons are both known as nucleons—particles found in a nucleus. The number of protons in the nucleus of the atom determines the element of the atom. Every helium atom, for example, will have exactly two protons.

However, some atoms of the same element can have different numbers of neutrons. These different forms that an element can take are called isotopes. Isotopes are often known by the number of total nucleons they have. For example, most helium in nature is of the isotope helium-4.[2] This means each atom has two protons and two neutrons. A small percentage of the helium in the world, though, is of the isotope helium-3. Helium-3 has two protons but only one neutron.

Some isotopes are unstable, meaning that over time they spontaneously change how they're arranged. They might change into a different isotope or even into a different element. These isotopes have nuclei where the protons and neutrons aren't as strongly bound together as they might be. As a result, a change in how the nucleus is organized eventually occurs.

The half-life of a radioactive isotope is the time it takes for half of it to decay. It's impossible to know when any particular atom of a radioactive isotope will decay. However, atoms of a given radioactive isotope decay at a predictable rate.[3] For example, radium-226 has a half-life of 1,600 years. That means that if you have a million atoms of radium-226, after 1,600 years about half a million of them will have decayed. You just won't be able to predict which ones!

Alpha Decay

In alpha decay, the radioactive element emits an alpha particle from its nucleus. For example, the most common isotope of uranium is called uranium-238. It has 92 protons and 146

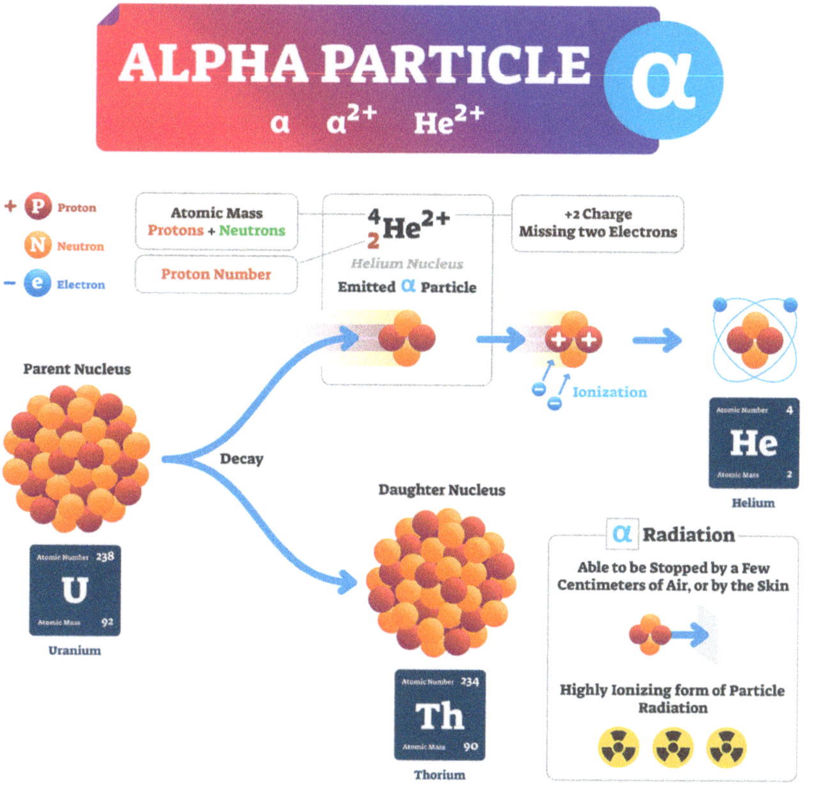

This diagram demonstrates alpha decay. The parent isotope emits an alpha particle and yields a new isotope.

neutrons. It's also radioactive. It will eventually emit an alpha particle that has two protons and two neutrons. Losing that particle means that the uranium-238 atom will become a thorium-234 atom (an isotope of thorium with 90 protons and 144 neutrons). This thorium-234 atom is known as the "daughter isotope."

Beta Decay

Beta decay is a little bit different. Generally in beta decay, a neutron turns into a proton and an electron is released along with a tiny amount of antimatter called an antineutrino. However, sometimes a proton turns into a neutron, releasing a positron and a tiny particle called a neutrino.[4]

Gamma Decay

Unlike alpha decay and beta decay, gamma decay doesn't change the number of protons or neutrons in an atom. Instead, it involves a change in how those nucleons are arranged.[5] These different arrangements of nucleons are called nuclear isomers. In gamma decay, an unstable isomer releases energy (a gamma ray) and forms a more stable isomer.[6]

One isotope that undergoes gamma decay is a particular isomer of cobalt-60. It has 27 protons and 33 neutrons. Once it decays, it will still have 27 protons and 33 neutrons, but those protons and neutrons will be arranged in a lower-energy state.

Fission and Fusion

Radioactive decay isn't the only kind of nuclear reaction that can produce gamma radiation. The powerful reactions known as fission and fusion also involve the release of gamma rays.

Nuclear Fission

Nuclear fission is the process by which an unstable atomic nucleus splits apart. This can happen in many ways, but the result is usually at least two smaller nuclei, one or more neutrons, and a lot of energy. Some of that energy is electromagnetic energy in the form of gamma rays.[7]

One key fact about nuclear fission is that it can be part of a chain reaction. The neutrons that result when a radioactive

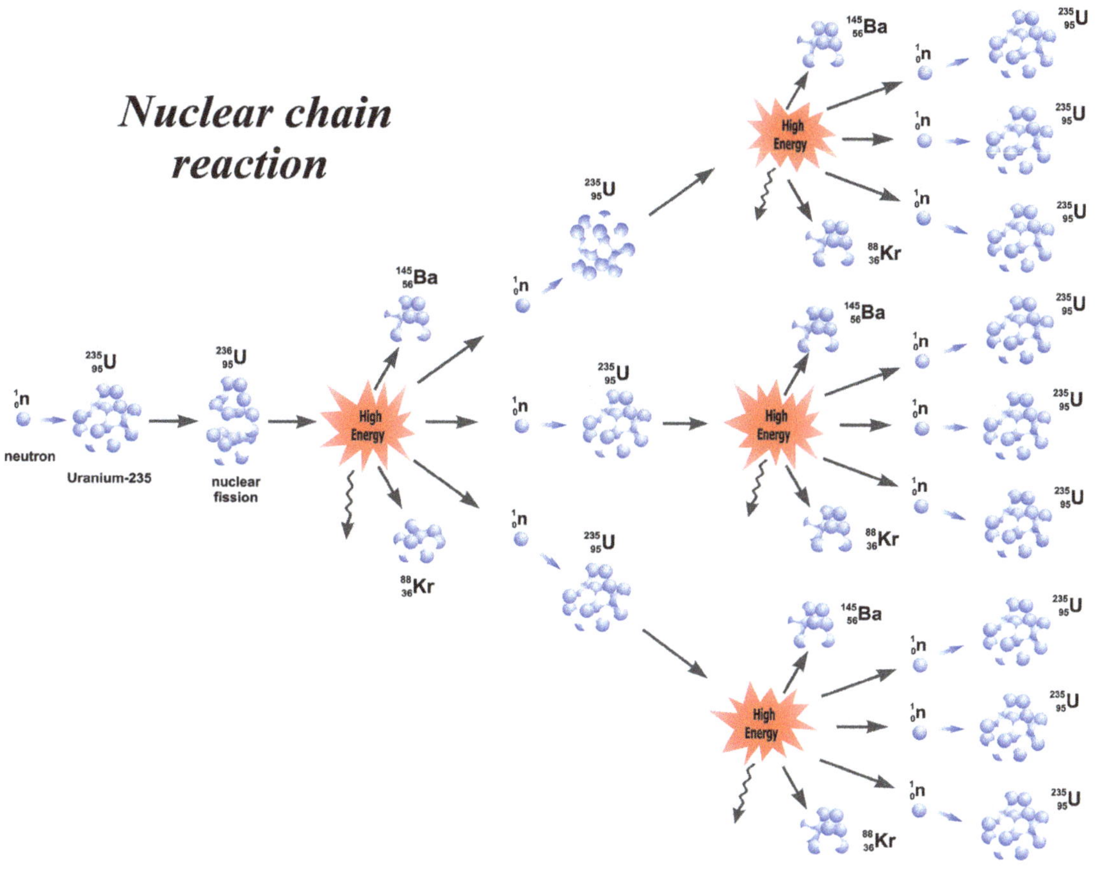

A controlled nuclear chain reaction can generate nuclear power. An uncontrolled nuclear chain reaction can create an explosion.

nucleus breaks apart can in turn break apart the nuclei of radioactive atoms around them. Those nuclei release their own neutrons, which continue the process. This means that once fission starts, it can release energy from many atoms in a short time. Controlled, this process can generate nuclear power. Uncontrolled, it can create nuclear explosions and the devastation of nuclear bombs.[8]

Nuclear Fusion

Nuclear fusion is the opposite of nuclear fission. It's the process by which two atomic nuclei fuse together with the result of a single larger nucleus. Like fission, this process can release a huge amount of energy, including energy carried by gamma rays. In fact, nuclear fusion is what powers the sun and other stars.[9]

GAMMA WEAPONS

Existing nuclear weapons usually work by releasing energy through nuclear fission or nuclear fusion. A gamma ray bomb would work a little differently. It would artificially trigger gamma decay in an unstable isomer, causing many atoms to release gamma rays at once. Unlike fission bombs, which require a large amount of material to create a nuclear chain reaction, a gamma bomb could be almost any size. That means it could add devastating power to both small and large missiles.[10] This power and flexibility make gamma weapons very desirable for national militaries—but also extremely dangerous in the wrong hands.

Stars have powerful gravitational forces in them. This gravitational force fuses isotopes of hydrogen (carrying a single proton) into helium (which has two protons).[11] Helium isotopes can be fused further into heavier and heavier elements.

The heaviest element that can be created by nuclear fusion is iron.[12] But stars are also the source of other, heavier elements. These elements are actually created through beta decay ... and exploding stars!

An atom of an element found in a star, such as iron, may pick up stray neutrons during the massive explosion known as a supernova. The enormous force of the explosion and the large

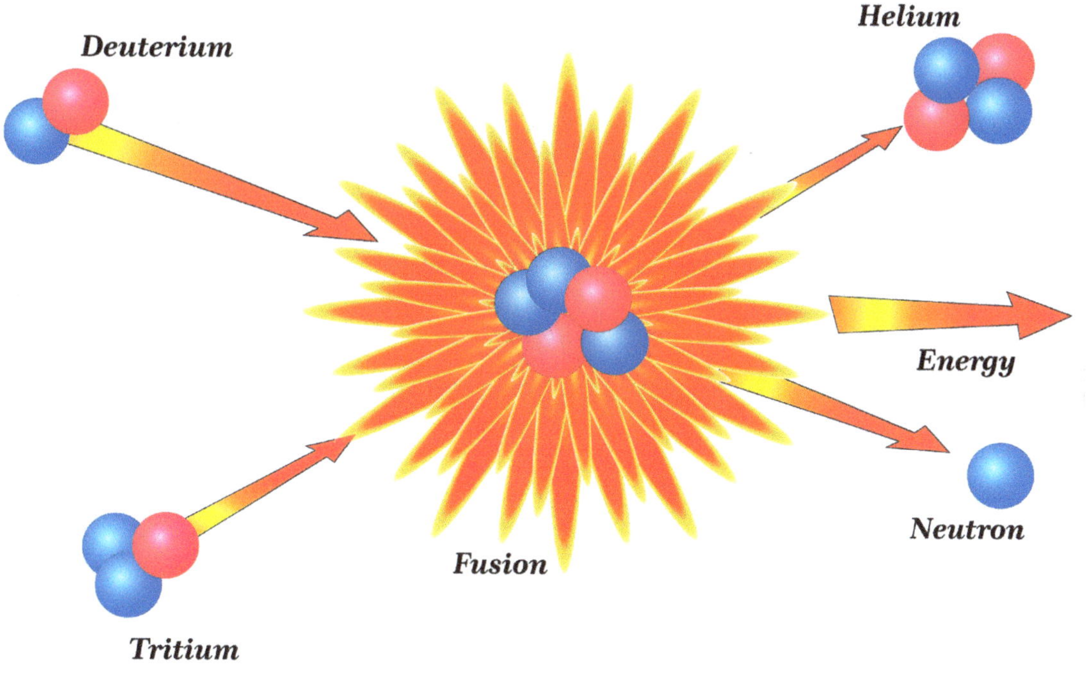

Nuclear fusion is the opposite of nuclear fission. A nuclear fusion reaction between tritium and deuterium is being explored as a way to generate nuclear power.

number of neutrons given off during it can create very heavy isotopes. If one of those isotopes undergoes beta decay, one of its neutrons turns into a proton and the atom becomes a new element.[13] For example, an isotope of iron (with 26 protons) may transform into cobalt (with 27 protons) and be spread across the universe by the supernova. This process gives rise to all the elements found in nature.[14]

Today's nuclear power is generally based on nuclear fission, but fusion power is a long-time goal of energy scientists.[15] It's possible to create a controlled fusion reaction using isotopes of hydrogen called deuterium (hydrogen-2) and tritium (hydrogen-3). The result is a form of helium, a neutron, and energy, some of which is released as gamma rays.

GAMMA RAY DETECTIVES

CHAPTER THREE

Gamma rays are so powerful you might think discovering them must have been easy. Actually, they weren't discovered until 1900 during a period of rapid scientific discovery sometimes called "the second scientific revolution."[1] Today, scientists use a variety of techniques to detect gamma rays, all beginning with that exciting period of discovery.

The Discovery of Gamma Rays

In 1895, Wilhelm Röntgen discovered X-rays, though at the time he didn't understand much about what they were.[2] Further discoveries were made by Henri Becquerel, who was studying the properties of X-rays. Becquerel thought that uranium might emit energy by absorbing it from the sun. However, he managed to prove otherwise, revealing that uranium somehow generated its own energy.[3] But it wasn't until the famous scientist Marie Curie began to study uranium with her husband, Pierre, that an atomic theory of radioactivity was proposed. Curie suggested that it was uranium's atomic structure that caused its strange energy emissions, and she coined the term "radioactivity" to describe the phenomenon.[4]

In 1899, Ernest Rutherford showed through experiments that radioactive elements emitted more than one kind of radiation.[5] One kind of radiation seemed to penetrate solid matter much better than the other. He called the weaker radiation alpha radiation and the stronger radiation beta radiation. Further investigation revealed that alpha particles were heavy and had a positive charge. It is now known they are a positively charged ion of helium. Meanwhile, beta particles were light and negatively charged. People now know they are electrons.

Ernest Rutherford (right) and Hans Wilhelm Geiger. Geiger invented the first working detector of individual alpha particles.

Curie and Rutherford were prominent scientists of their day. But it was a relatively unknown scientist named Paul Villard who in 1900 found there was a third type of radioactivity.[6]

Like the Curies, Villard was working in Paris. He was very interested in Marie Curie's discovery of radium as well as Roentgen's discovery of X-rays. He discovered a third kind of radiation present in radioactive materials. These rays were much more penetrating than alpha particles. They could even pierce lead foil. But unlike beta particles, they weren't affected by magnetic fields. They had to be something different, something more like X-rays than beta particles. They were named gamma rays, and in 1914 it was Rutherford who determined that they were a form of light, like X-rays but with shorter wavelengths.[7]

GAMMA ARMOR

Gamma rays can be really dangerous to humans, but that doesn't mean people just ignore them. Sometimes people are exposed to gamma rays through their jobs—especially scientists. So how do they protect themselves?

The three best strategies are time, distance, and shielding. The longer you are exposed to a radioactive substance, the more it will affect you. So limiting the amount of time you spend in the vicinity of such a substance is an important strategy. Distance is also important. The farther you are from a source, the less it will affect you. Shielding can also help. The intensity of radiation is lessened as it passes through matter. The best shields against gamma rays are large quantities of water, concrete, and lead.[9]

Detecting Gamma Rays

So now we know who first discovered gamma rays. But how are they detected if they can't be seen? Trying to detect and measure gamma rays can be tricky. Gamma ray wavelengths are so short that they might pass between the atoms of a detector without being noticed.[8] And unlike visible light, they are ionizing radiation. That means that when they interact with matter, they may lose much or all of their energy interacting with the electrons in atoms. As a result, it's hard to capture a gamma ray in a mirror or lens with any accuracy.

Scientists have developed specialized gamma ray detectors that use the gamma rays' particular properties to reveal their presence. Gamma rays' habit of ionizing matter is a problem, but it can also be a solution. If gamma rays tend to knock electrons loose from atoms, then rather than trying to measure gamma rays directly, you can instead measure the electrons they send flying. There are many ways to use ionization to detect and measure gamma rays, but two of the most common are called scintillation detectors and solid state detectors.

Scintillation Detectors

The word "scintillation" refers to a small sparkle of light. In a scintillation detector, ionizing radiation is turned to scintillations, or bursts of visible light.

In a scintillation detector, gamma rays interact with a luminescent material (a material that emits light without being heated). The material is often a crystal or other dense substance. When the gamma rays interact with this material, its atoms

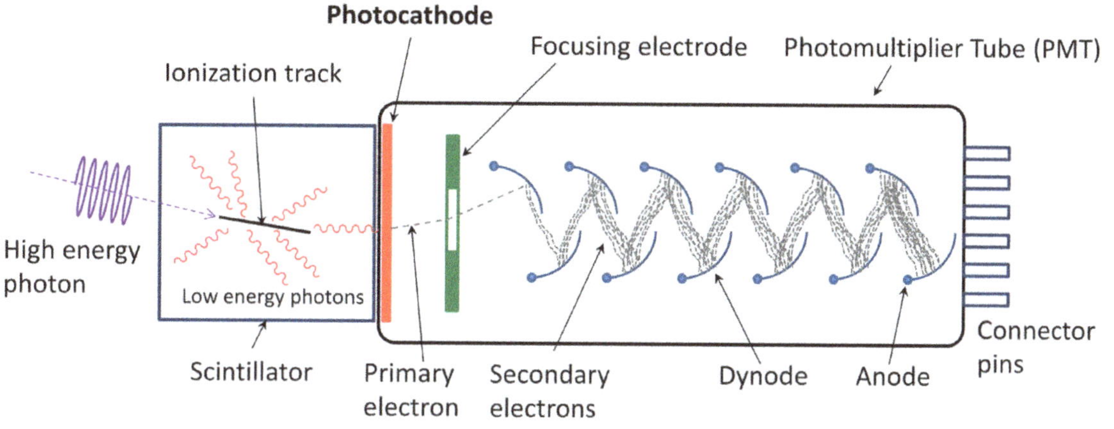

This diagram shows how a photon travels through a scintillation detector.

scintillate, releasing photons. These photons are then typically captured by a photomultiplier tube, a device that turns weak light signals into detectable electrical signals.

The capture begins with a photocathode, which is a device that emits electrons when illuminated. These electrons are passed through a photomultiplier tube, which increases the number of electrons so that they can be more easily detected. The result is an output of electrons that's proportional to the strength of the gamma ray.[10]

Solid-State Detectors

Solid-state detectors work a little differently than scintillation detectors. Instead of a scintillation material, there is a "solid-state" semiconductor crystal. This semiconductor crystal has a special sensitive region called the depleted region. The depleted region is prepared in such a way that when gamma rays interact with it, the free-moving electrons produced build a charge, which is detected directly.

The result is a detector that is more precise than a scintillation detector. However, it has downsides. Solid-state detectors are more likely to degrade over time.[11]

Gamma Imaging

These relatively simple detectors collect gamma rays and output readings about their strength. However, there are also machines that are used to create images based on gamma rays. Gamma imaging is very useful to both scientists using gamma rays to study the skies and doctors using them for medical scans.

In either case, the gamma imaging device (sometimes called a gamma camera) uses a detector or array of detectors that transmit signals to a computer. The computer then uses mathematical principles to attempt to reconstruct where each gamma photon was released from. This forms an image of the radioactive material emitting the gamma rays. That can be a massive event in space or a test material filling a patient's lungs.[12]

CHAPTER FOUR

GAMMA RAYS ON EARTH

There are radioactive materials all around us on planet Earth. Luckily, most of the radiation people encounter day to day is too weak to do us real harm. Radiation around us can come from within the ground, from the skies, or even from within our own bodies.

Some of the radioactive material on Earth is natural and was discovered by humans. Others are synthetic, made by humans, intentionally or unintentionally.

Natural Radioactive Materials

Like other heavy elements, many radioactive elements were created in stars and supernovas billions of years ago, before Earth was formed. When all those elements came together to create planet Earth, those radioactive materials ended up in Earth's crust.[1] Some of those materials, called primordial isotopes, have half-lives so long that they still exist today.

One example is uranium-238. Uranium-238 is the most common isotope of uranium on Earth, with a half-life of over four billion years. Uranium-238 is so common on Earth that it produces about 40 percent of all radioactive heat within the planet.[2]

Another important primordial isotope is potassium-40. Potassium-40 is one of several naturally occurring radionuclides found in the human body! A small percentage of all potassium found in nature is potassium-40. It undergoes beta decay and is the major source of gamma radiation that can escape the human body. Because potassium-40 is found naturally in all potassium, potassium-rich foods such as bananas and Brazil nuts also tend to be unusually radioactive.[3] The good news is the radiation is so weak that the potassium will probably do you more good than harm. Don't toss your bananas in the trash just yet!

As primordial isotopes decay, some of them transmute into other elements. When a primordial isotope decays into another radioactive isotope, the resulting isotope is called a secondary radionuclide. Secondary radionuclides generally have much shorter half-lives than their parent isotopes.[4]

When uranium-238 undergoes alpha decay, it transmutes into thorium-234. Thorium-234 is itself radioactive and undergoes beta decay to transmute into protractinium-234. Protractinim-234 is also radioactive! In fact, Uranium-238 is the beginning of a long decay chain, which includes isotopes of uranium, radium, polonium, and radon, and others. The chain eventually ends with a stable isotope of lead.[5] These secondary radionuclides have half-lives of as long as 245,000 years or as little as 0.00016 seconds! Some of the isotopes in this chain undergo gamma decay as well as alpha or beta decay, so uranium-238 is a source of a mix of alpha, beta, and gamma radiation.[6]

Not all natural radioactive material has its origins in supernovas. Some are cosmogenic. This means they are created by cosmic rays interacting with elements on Earth. One example of this is carbon-14, an isotope of carbon that scientists often use to help date fossils. Another is sodium-24, an important source of gamma rays used in nuclear medicine. Cosmogenic radionuclides are produced not only on Earth, but also on other planets—and scientists believe they can learn more about other planets based on the gamma rays they emit.

Human-Made Radioactive Materials

Since the dawn of nuclear science, humans have created a number of radioactive isotopes. Some of them are created on purpose. Others are

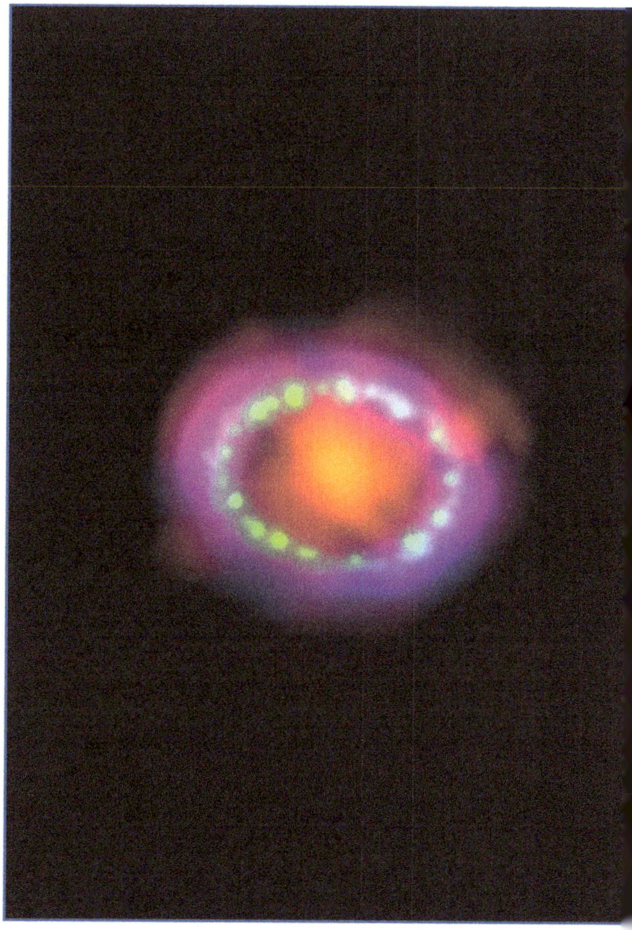

Many radioactive elements are the product of supernovas. This composite image uses data from several different telescopes to show what a supernova looks like.

a side effect of the work done with nuclear materials.

Scientists have never discovered large amounts of any element with an atomic number higher than that of uranium (atomic number 92) naturally in the universe.[8] However, they have created elements with higher atomic numbers. Scientists can generate new elements by blasting existing elements with particles in a device called a particle accelerator.[9] Sometimes these particles are other elements, and the resulting fusion creates a new, heavier element. Moscovium is an element of atomic number 115, meaning it has 115 protons. Scientists generated it by fusing americium (atomic number 95) with calcium (atomic number 20).[10]

Other times, an element may be placed into a nuclear reactor, where it is exposed to neutron radiation. The element in the reactor captures additional neutrons, which later change to protons due to beta decay.[11] Plutonium, for example, has an atomic number

TERRESTRIAL GAMMA FLASHES

Lightning can also produce gamma radiation in a phenomenon known as a terrestrial gamma-ray flash. Terrestrial gamma-ray flashes happen very quickly and are hard to detect. They last less than a millisecond. Lightning happens when a cloud builds up electrical charge and this charge is attracted to an opposing charge elsewhere in the clouds or in the ground. In a terrestrial gamma-ray flash, the electric field in the lightning becomes strong enough to accelerate electrons to near the speed of light. When these electrons crash into air molecules, they release a burst of gamma rays.[7]

of 94, two higher than that of uranium. It is sometimes produced in nuclear reactors when uranium-238 picks up an extra neutron to become uranium-239. Uranium-239 undergoes beta decay to become neptunium-239 and then a second beta decay to become plutonium-239.

All elements heavier than uranium are unstable and decay over time, many of them releasing gamma rays as they do so.

However, not all human-made radioactive isotopes are heavy elements not found in nature. Scientists can create isotopes and isomers of abundant elements, such as zinc, sodium, and

The Large Hadron Collider, pictured here, is the world's largest particle accelerator.

cobalt, that emit gamma radiation. People also sometimes create radioactive isotopes accidentally. Nuclear waste from reactors and nuclear fallout from atomic weapons, called fission products, often emit neutron radiation. Those neutrons may attach themselves to stable elements, creating unstable isotopes that emit radiation themselves. This is one of the reasons that radioactive waste is so dangerous.[12]

GAMMA RAYS IN SPACE

CHAPTER FIVE

Like other kinds of electromagnetic radiation, gamma rays are found throughout the universe. If you look into the sky and see something shining down, there's a good chance there are gamma rays shining down along with the visible light.

Stars, Supernovae, and More

Gamma rays have wavelengths much shorter than those of visible light so they're invisible to human eyes. But if people could see them, the night sky might reveal much more than just the stars and planets seen today. This is why scientists have developed gamma-ray telescopes. These telescopes include devices, such as scintillation detectors, that help capture gamma rays and create images of them. Scientists use them to observe some of the most spectacular and energetic phenomena in the universe.

Stars

Stars emit gamma rays as part of the usual fusion process that powers them. But some stars produce more than others.[1]

Any star may undergo an unusual release of gamma rays as part of a solar flare. Stars have an atmosphere, and magnetic energy can build up in that atmosphere until it's suddenly released in a flare of energy across the electromagnetic spectrum. This solar flare releases something close to ten million times the energy of a volcanic eruption.[2] But even that massive amount is a tiny fraction of a star like the sun's daily output.[3]

Scientists also use gamma radiation to study some stars very unlike our own. White dwarf stars have about as much mass as

This telescope in the Canary Islands detects Cerenkov radiation, a type of radiation caused by gamma rays.

the sun, but are about the size of Earth, so they're extremely dense. They are nearly a million times as dense as water. White dwarf stars tend not to be very bright, but they've been linked to massive eruptions of gamma rays due to novae on their surface.

Neutron stars are even more dark and dense than white dwarfs. Typically, a neutron star has a mass greater than that of the sun condensed into a ball about 12 miles (19.3 km) across. Their density is something like a hundred trillion times that of water. They are, in fact, so dense that their density is similar to that of a nucleus of an atom. They're thought to be composed mostly of neutrons, so one way you could think of a neutron star is as the nucleus of a single atom twelve miles across!

A kind of fast-rotating neutron star called a pulsar emits massive amounts of gamma radiation. Neutron stars are also thought to be responsible for short gamma-ray bursts. More on them later.

Beyond the Stars

Many other phenomena can be studied by the gamma rays they emit. Some, like supernova remnants and black holes, are the remains of dead stars.

A supernova is an enormous explosion that sends matter and energy from a star flying across the universe. Supernovae release massive amounts of radiation, including gamma radiation. They also create massive nebulae that expand across galaxies, filled with freshly generated radioactive elements. Supernova remnants also contain shockwaves that create threads of radioactive materials, some of which release gamma rays. Studying supernovae and their remnants helps scientists understand how galaxies form and the role stars and supernovae have in spreading elements across the universe.

A black hole is an astronomical object with such an intense gravitational pull that nothing, not even light, can escape if it comes too close. Black holes make up the cores of galaxies and are responsible for the formation of quasars. Quasars are extremely bright objects found at the center of galaxies. They are formed by gas pouring into black holes at high speeds. A kind of quasar called a blazar appears particularly bright with gamma rays because of its orientation toward Earth.[4]

Less spectacular, but fascinating for anyone who's ever dreamed of setting foot on another world, is gamma-ray spectroscopy, the study of the spectrum of gamma rays emitted by different atoms. Scientists use a device called a gamma-ray spectrometer to study the gamma rays emitted by atoms when they're struck by cosmic rays—including atoms on other planets. NASA's Mercury Surface, Space Environment, Geochemistry, and Ranging (MESSENGER) program studied the surface of the planet Mercury using gamma rays. Different substances emit different gamma rays when struck by cosmic rays. This makes it possible to tell apart elements like hydrogen, oxygen, and iron by studying the gamma rays emitted by a planet. This helps us figure out what the surfaces of distant planets look like.[6]

Gamma-Ray Bursts

Stars and other astronomical bodies may shine in the gamma portion of the spectrum, but in all the universe nothing since the Big Bang has matched the energetic power of the gamma-ray burst. A gamma-ray burst, also called a GRB, can release more

FERMI CONSTELLATIONS

NASA's Fermi Gamma-Ray Space Telescope has found thousands of gamma ray–emitting objects in space. This is comparable to the number of stars used to make up the constellations in the night sky. For fun, the Fermi team decided to make up some gamma-ray constellations. They include Godzilla, the Hulk, the Eiffel Tower, and Albert Einstein. Of course the gamma objects Fermi finds usually aren't stars, and you can't see them unless you have a gamma-ray telescope. But it's fun to pretend![5]

NASA's Fermi Gamma-Ray Space Telescope

energy in ten seconds than a star like our sun may release in its entire lifetime.[7]

A GRB can last anywhere from a tiny fraction of a second to a few minutes. Scientists have divided them into two classes: short, with an average duration of about three-tenths of a second, and long, which have an average duration of about thirty seconds. These blasts of energy occur when stars collapse, and they are more powerful even than supernovae.[8]

Short gamma-ray bursts are believed to occur when neutron stars collide, either with each other or with a black hole. When two neutron stars merge and collapse into a black hole, the massive destruction releases a flash of gamma rays.

Long gamma-ray bursts are associated with a kind of stellar explosion even more powerful than a supernova called a hypernova. While a supernova occurs when a star explodes, a hypernova occurs when the core of an enormous star, at least thirty times larger than our own sun, collapses into a black hole so quickly that an enormous void opens up within the star. If the star is spinning rapidly, the collapse of the rest of the star into the black hole can whip superheated plasma into twin jets of focused particles and energy. Not every hypernova results in GRBs, at least not ones visible from Earth. But GRBs from hypernovae are the most energetic events in the universe. The most powerful one we've observed was 2.5 million times brighter than the brightest known supernova.[9]

Their immense power makes GRBs visible even over great distances. The most distant object humans have observed is a gamma burst that occurred in the early days of the universe and traveled for billions of years to be seen by us. But their power also makes them incredibly dangerous. If a GRB pointed at Earth erupted too close to us, it could potentially annihilate all life on the planet. In fact, some scientists suspect that a mass extinction event 450 million years ago, known as the Ordovician mass extinction, may have been triggered by a GRB.[10] Fortunately, GRBs are very rare. There are often millions of years between GRBs in any given galaxy.

GAMMA RAYS AT WORK

Electromagnetic radiation is a powerful tool that has applications in virtually every industry. Gamma rays are no exception. They are used every day in engineering and agriculture.

Industrial Applications of Gamma Rays

Gamma rays aren't only useful for astronomers. They have uses in a variety of industries, from mining to the auto industry. Here are just a few of the ways gamma rays shape today's society.

Nondestructive Testing

Nondestructive testing is the process of checking the quality of a product without damaging it. Imagine that you want to test the wheels of a car before it leaves the factory to ensure that they won't unexpectedly break when the car is driven. What's the best way to perform that test without adding wear and tear to the car, making it more likely to break in the future?

Well, one way to do a test like that is to use radiation—gamma rays or X-rays, usually—to make a highly detailed image of the wheels. Gamma ray imaging can reveal imperfections that might not be visible to the human eye, particularly imperfections in the thickness or composition of welds that hold pieces together. Radiation testing is used on everything from aircraft turbines to oil pipelines to ski-lift cables.[1]

As another example, an isotope that emits gamma rays can be added to a liquid or gas. That liquid or gas can then be pumped through a closed system of tubes. By tracking how the

This truck is being scanned by the VACIS II system, which uses a gamma ray imaging technique to examine the contents.

radiation spreads through the system, an engineer can check for leaks or blockages in the tubes.

Radiation Crosslinking

A polymer is a kind of large molecule consisting of small, similar groups of atoms linked together. Polymers are associated with plastics, resins, and other useful materials. In recent years, scientists have developed a process called crosslinking that uses radiation to improve the quality of these materials.[2]

Essentially, carefully controlled radiation breaks down molecular bonds within polymers and creates new ones. This makes a kind of network of bonds that gives the polymer new properties. Materials treated in this way may be stronger and more resistant to things like heat, friction, or chemical damage.[3]

Geology and Exploration

Gamma-ray spectrometry is used in geology to explore Earth and better understand the composition of its surface. Gamma rays are particularly good for this kind of exploration because of how penetrating they are. Alpha radiation and beta radiation travel poorly even through open air.

Geologists use gamma rays in many kinds of exploration. They use gamma-ray spectrometers from planes to analyze large areas of land. They use gamma-ray spectrometers dug into the ground to figure out the makeup of rock they're drilling into. They even use handheld gamma spectrometers to measure gamma radiation on the ground.[4]

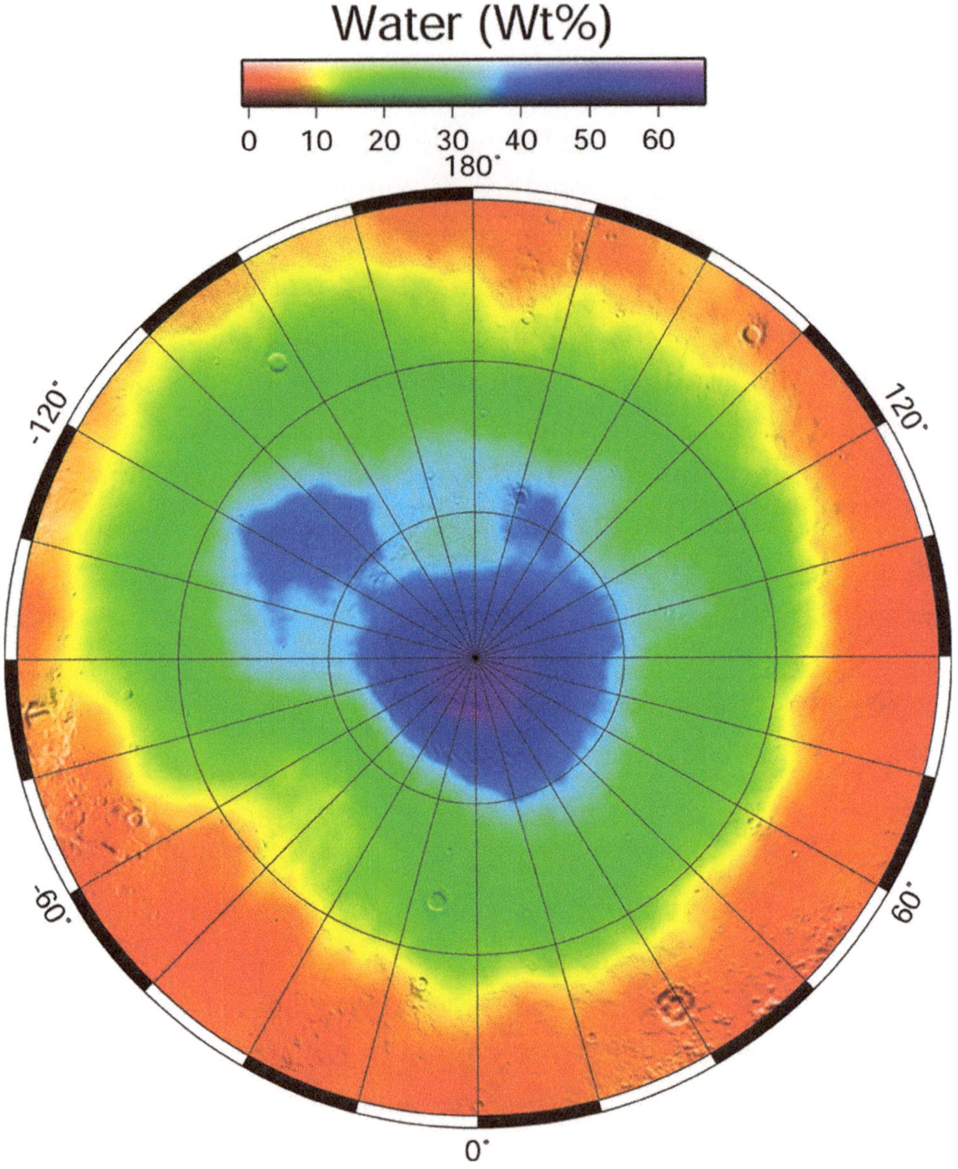

This map shows water by weight in materials near the north pole of Mars. The data was collected by a gamma-ray spectrometer on NASA's Mars Odyssey spacecraft.

The levels of gamma radiation geologists find can tell them about the composition of the area. Different kinds of rock have different chemical compositions. In particular, they have different levels of radioactive isotopes of potassium, uranium, and thorium. By measuring the gamma emissions in an area of rock, scientists can figure out the amount of these isotopes in the area. This tells them what kinds of minerals they're likely to find there.

Gamma Radiation in Agriculture

Gamma radiation may be deadly, but those same deadly properties are used to help humanity survive. Agriculture is the industry of creating food and other products from plants and animals. Gamma radiation is used extensively in agriculture to make food safe to consume, to protect it from pests, and even to create better food crops.

GEMSTONE IRRADIATION

Another thing gamma rays are used on is gemstones. Gamma exposure can actually cause gemstones to change colors! The color change depends on how the electrons reorganize with gamma radiation. Topaz in its purest form has no color, but with gamma radiation it can change to a brown or even a bright blue. In fact, almost all the blue topaz in the world market was made blue with radiation. Quartz can turn into amethyst with gamma radiation, pearls turn blue or grey, and even diamonds can be turned yellow or green.[5]

Sterilization

One of the key ways that radiation is used in the production and distribution of food is sterilization. Food can become dangerous to consume because of contamination by microorganisms. Some dangerous microorganisms include bacteria (you may have heard of *E. coli* that can cause foodborne illness), fungi (mold, for example), and parasites (like the tiny worm that causes trichinosis). Sterilization destroys these organisms before they can make food dangerous to eat.

These oysters are being prepared for shipment to a facility that will use gamma rays to make them safer to eat.

Gamma rays are one type of radiation used to sterilize food. Radiation emitted by cobalt-60 and cesium-137 are often used for this purpose. The radiation kills these microorganisms without making the food itself dangerously radioactive.[6]

Irradiating food can have other benefits as well. It can prevent spoilage by destroying microorganisms that cause food to go bad, sometimes extending the shelf life of food for years. It can destroy insects or their eggs, stopping them from ruining tropical fruit. It can even slow down how quickly food ripens to keep it fresh longer.

Sterile Insect Technique

Gamma rays aren't used only to kill insects in food. They can also be used to control the populations of pest insects. Fewer pests mean less food lost and more food brought to people's tables.

The sterile insect technique is a method of controlling insect populations that doesn't involve killing them directly. Instead, scientists actually breed the insects in massive numbers! Why would they want more pests? Because they use gamma radiation to sterilize all the male insects. In this case, sterilize doesn't mean kill. It means making the insects unable to reproduce.

The scientists then release the sterilized male insects. These sterilized insects compete with the existing male insects for the chance to reproduce with female insects. Many of the existing male insects are crowded out, reducing the total number of insects in future generations.[7]

This method of pest control has some advantages. The sterile insects eventually die out, so they don't become a permanent part of the environment. And because it's targeted at a single species, it won't accidentally poison other organisms as pesticides might.

Mutation Breeding

Sometimes, gamma rays are used to cause mutations. A mutation is a change in an organism's DNA that can be passed on to its offspring. Gamma radiation, like other ionizing radiation, can have an effect on living creatures' DNA. Usually the changes are harmful, and they raise the risk of cancer.[8] But occasionally a mutation to something like a plant's seed may help rather than hurt.

Scientists sometimes irradiate seeds and then look for beneficial mutations in the plants that result. These mutations may be resistance to disease or weather, or an increase in the amount of food they produce. When a useful mutation is found, the resulting crop can be bred and the useful mutation spread.[9]

CHAPTER SEVEN

GAMMA RAYS IN MEDICINE

Some of the most important uses of radiation are in the field of medicine. It may seem strange that deadly radiation can be used to heal. But major advancements in medical diagnosis and treatment have depended on gamma rays. In fact, according to the Center for Nuclear Science and Technology Information, about one-third of all hospital treatments involve a kind of radioactivity.[1]

Gamma Ray Imaging

The use of radiation to penetrate the body and form an image is called diagnostic radiology. A diagnosis is the identification of an illness or injury. Doctors and other technicians who use these techniques are called diagnostic radiologists.

The most famous use of diagnostic radiology might be the X-ray images that allow medical specialists to look at a patient's insides without invasive surgery. In an X-ray, radiation passes through the patient's body and is detected on the other side. Dense materials in the body, like bone, absorb some of the X-ray radiation. The X-ray detector creates an image that distinguishes the harder, denser objects in the body from the softer, less dense ones.

Today, diagnostic radiology goes beyond X-rays. The science of nuclear medicine involves injecting small amounts of radioactive isotopes into the body and using their emissions for diagnostic imaging. One common example of nuclear medicine using gamma rays is positron emission tomography (PET). PET scans create three-dimensional images of parts of the body,

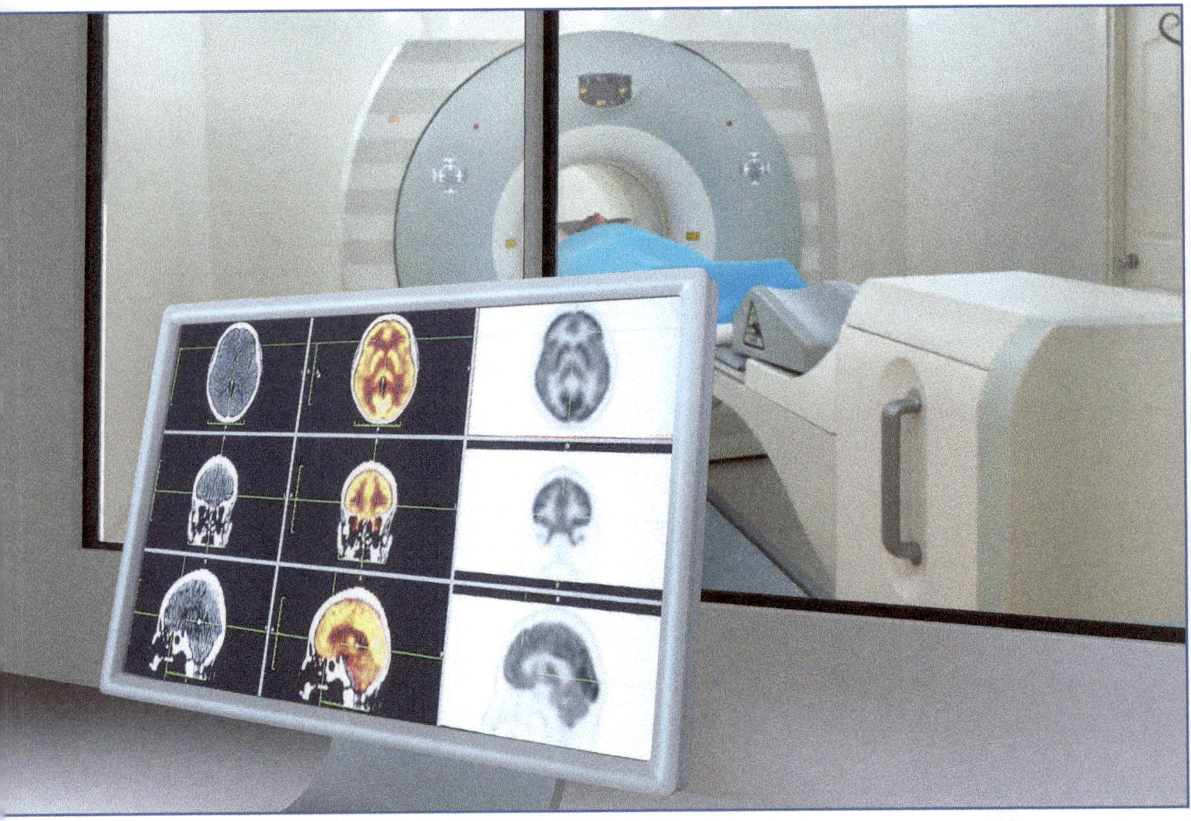

Medical scanners use various types of radiation to help diagnose patients. Some of them use gamma radiation.

particularly the brain and the heart. It gives doctors a good idea of how some bodily processes are working.

A PET scan involves injecting the patient with an isotope that emits positrons. These positrons quickly meet electrons in a process called pair annihilation. In pair annihilation, a positron and an electron meet and destroy each other, emitting a pair of gamma rays. A gamma-ray detector and a computer can then be used to figure out where the positrons were emitted and generate an image of the biological processes in the targeted area. This technique can help diagnose cancer, help doctors understand the workings of the heart, and even diagnose mental problems by imaging the brain.[2]

Gamma Sterilization

Just as it does with food, the deadly power of gamma radiation can be used to make medical equipment safe. Microorganisms such as those that can contaminate food can be a serious danger to patients in medical settings. It's vital that surgical equipment, needles, and tissues for graft or transplant are sterilized before they come into contact with a patient to prevent serious infections.

Gamma irradiation generally uses a particular isotope, cobalt-60, to generate gamma rays. This method of irradiation has many benefits compared to other forms of sterilization. It's a very quick method of sterilization. It is very penetrating radiation, so it can sterilize well below the surface of most objects. It can even sterilize objects that have been packaged and sealed.[3]

However, research shows that there may be serious side effects

GAMMA KNIFE SURGERY

A very specialized way gamma rays are used in medicine is in gamma knife radiosurgery. This name is a bit tricky because there is actually no knife and no "surgery" involved. The procedure uses about 200 highly focused gamma ray beams rather than a scalpel. Because the beams are so focused, healthy tissue isn't affected. Gamma knife radiosurgery is used to treat problems in the brain such as brain tumors, nerve problems, epilepsy, and malformations of veins and nerves. It can usually be done in one treatment in one day, and the patient doesn't feel or hear anything while the gamma rays are working.[4]

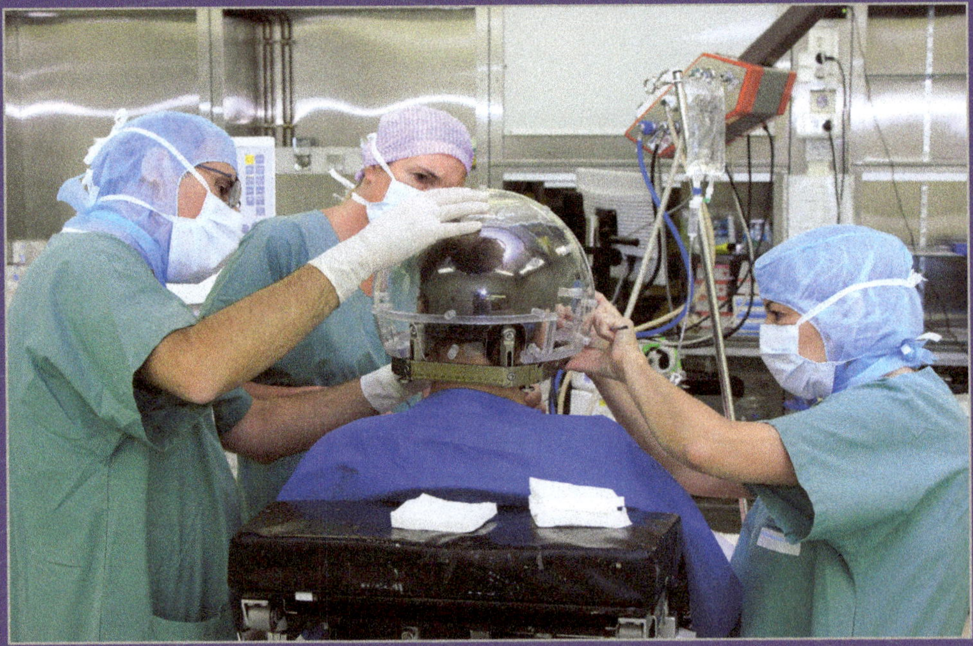

Doctors in France use gamma knife surgery to treat a tumor in a patient.

to gamma sterilization. The power of gamma rays isn't always easy to control. Gamma radiation can weaken some polymers it's exposed to, especially with repeat exposure. Gamma radiation can also weaken tissues used in grafts.[5]

Whether used for imaging, sterilization, or surgery, gamma radiation can also pose risks for the people involved. How much radiation a person is exposed to is often measured in units called grays and sieverts.[6] Grays measure the amount of radiation a body is exposed to. But not all radiation interacts with the body the same way. Sieverts measure the effective amount of radiation a body is exposed to. Governments and hospitals have regulations controlling how many milligray or millisieverts of radiation patients and workers can be exposed to.[7]

Like scientists who work with radiation, medical professionals use the principles of time, distance, and shielding to minimize exposure to radiation. Patients undergoing tests often wear lead aprons to protect sensitive body parts during imaging. Radiation exposure for patients is usually low. But technicians work with radiation again and again, and so they must take extra precautions so that repeat exposure doesn't cause problems. They often step away and behind special barriers while tests are run to ensure their exposure is minimal.

Gamma rays have tremendous power to help us learn more about ourselves, our world, and the universe. But it's important to take the utmost care when using that power, or people can do as much harm as good.

CHAPTER NOTES

INTRODUCTION
THE ELECTROMAGNETIC SPECTRUM

1. *Encyclopaedia Britannica*, "Photon," accessed February 14, 2019, https://www.britannica.com/science/photon.
2. NASA Science, "Anatomy of an Electromagnetic Wave," accessed February 14, 2019, https://science.nasa.gov/ems/02_anatomy.
3. Ibid.
4. Boundless Physics, "The Electromagnetic Spectrum," accessed February 14, 2019, https://courses.lumenlearning.com/boundless-physics/chapter/the-electromagnetic-spectrum.

CHAPTER 1.
WHAT ARE GAMMA RAYS?

1. *Encyclopaedia Britannica,* "Gamma Ray," accessed February 14, 2019, https://www.britannica.com/science/gamma-ray.
2. Monmouth County, NJ, "Radiation Health Basics," accessed February 14, 2019, https://co.monmouth.nj.us/documents/118/RADIATION%20HEALTH%20BASICS.pdf.
3. *Encyclopaedia Britannica,* "Gamma Ray."
4. *Encyclopaedia Britannica,* "Radiation," accessed February 14, 2019, https://www.britannica.com/science/radiation.
5. Purdue University College of Science, "Atoms vs. Ions," accessed February 14, 2019, http://chemed.chem.purdue.edu/genchem/topicreview/bp/ch2/atom_ion.html.

6. World Health Organization, "What Is Ionizing Radiation?," accessed February 14, 2019, https://www.who.int/ionizing_radiation/about/what_is_ir/en/.
7. Mayo Clinic, "Radiation Sickness," accessed February 14, 2019, https://www.mayoclinic.org/diseases-conditions/radiation-sickness/symptoms-causes/syc-20377058.
8. *Encyclopaedia Britannica,* "Radiation."
9. *Encyclopaedia Britannica,* "Radioactivity," accessed February 14, 2019, https://www.britannica.com/science/radioactivity.
10. *Encyclopaedia Britannica,* "Discovery of Radioactivity," accessed February 14, 2019, https://www.britannica.com/science/atom/Discovery-of-radioactivity.
11. Mirion Technologies, "Types of Ionizing Radiation," accessed February 14, 2019, https://www.mirion.com/learning-center/radiation-safety-basics/types-of-ionizing-radiation.

CHAPTER 2.
THE SECRET ORIGINS OF GAMMA RAYS

1. Boston University Physics, "Radioactive Decay," accessed February 14, 2019, http://physics.bu.edu/py106/notes/RadioactiveDecay.html.
2. Jefferson Lab, "Isotopes of the Element Helium," accessed February 14, 2019, https://education.jlab.org/itselemental/iso002.html.
3. *Encyclopaedia Britannica*, "Half-life (radioactivity)," accessed February 14, 2019, https://www.britannica.com/science/half-life-radioactivity.

4. *Encyclopaedia Britannica*, "Beta Decay," accessed February 14, 2019, https://www.britannica.com/science/beta-decay.
5. *Encyclopaedia Britannica*, "Gamma Decay," accessed February 14, 2019, https://www.britannica.com/science /gamma-decay.
6. Ibid.
7. *Encyclopaedia Britannica*, "Nuclear Fission," accessed February 14, 2019, https://www.britannica.com/science/ nuclear-fission.
8. Atomic Heritage Foundation, "Science Behind the Atom Bomb," accessed February 14, 2019, https://www. atomicheritage.org/history/science-behind-atom-bomb.
9. *Encyclopaedia Britannica*, "Nuclear Fusion," accessed February 14, 2019, https://www.britannica.com/science/ nuclear-fusion.
10. David Hambling, "Gamma-ray Weapons Could Trigger Next Arms Race," *New Scientist*, August 13, 2003, https://www. newscientist.com/articledn4049-gamma-ray-weapons-could-trigger-next-arms-race.
11. *Encyclopaedia Britannica*, "Nuclear Fusion."
12. *Encyclopaedia Britannica*, "Nucleosythesis," accessed February 14, 2019, https://www.britannica.com/science/ nucleosynthesis.
13. Ibid.
14. Ibid.
15. Michael MacRae, "Cold Fusion 25 Years Later," ASME.org, May 2014, https://www.asme.org/engineering-topics/ articles/nuclear/cold-fusion-25-years-later.

CHAPTER 3.
GAMMA RAY DETECTIVES

1. Linda Hall Library, "Scientist of the Day - Paul Villard," accessed February 14, 2019, https://www.lindahall.org/paul-villard/.
2. Lawrence Berkeley National Laboratory, "The Discovery of Radioactivity," accessed February 14, 2019, https://www2.lbl.gov/abc/wallchart/chapters/03/4.html.
3. Ibid.
4. Ibid.
5. *Encyclopaedia Britannica*, "Discovery of Radioactivity," accessed February 14, 2019, https://www.britannica.com/science/atom/Discovery-of-radioactivity.
6. Leif Gerward, "Paul Villard and His Discovery of Gamma Rays," Physics in Perspective, December 1999, https://link.springer.com/article/10.1007/s000160050028.
7. Jim Lochner, "Ask an Astrophysicist," February 9, 1998, http://teacherlink.ed.usu.edu/tlnasa/reference/imaginedvd/files/imagine/docs/ask_astro/answers/980209c.html.
8. Nasa Science, "Gamma Rays," accessed February 14, 2019, https://science.nasa.gov/ems/12_gammarays.
9. Center for Nuclear Science and Technology Information, "Protecting Against Exposure," accessed 2/14/19, http://nuclearconnect.org/know-nuclear/science/protecting.
10. Hastings A. Smith, Jr. and Marcia Lucas, *Passive Nondestructive Assay of Nuclear Materials*, (Los Alamos, NM: Los Alamos National Laboratory, 1991), pp. 43–64, https://www.lanl.gov/orgs/n/n1/panda/00326398.pdf.

11. Smith, Jr. and Lucas, *Passive Nondestructive Assay of Nuclear Materials.*
12. TeacherLink, "How Do You Detect a Gamma-Ray?," Utah State University, accessed February 14, 2019, http://teacherlink.ed.usu.edu/tlnasa/reference/imaginedvd/files/imagine/docs/science/how_l1/gamma_detectors.html.

CHAPTER 4.
GAMMA RAYS ON EARTH

1. World Nuclear Association, "Naturally-Occurring Radioactive Materials," accessed February 14, 2019, http://www.world-nuclear.org/information-library/safety-and-security/radiation-and-health/naturally-occurring-radioactive-materials-norm.aspx.
2. Ricardo Arevalo Jr., W F W Mcdonough, and Mario Luong, "The K/U Ratio of the Silicate Earth: Insights into Mantle Composition, Structure and Thermal Evolution," *Earth and Planetary Science Letters* 278 (2009), https://pdfs.semanticscholar.org/425e/f444a981d1daf7ea98ab3467204ebe1e9613.pdf?_ga=2.13464693.331033830.1550211898-2055069150.1550211898/.
3. Anne Marie Helmenstine, PhD, "10 Radioactive Everyday Products," ThoughtCo., March 20 2018, https://www.thoughtco.com/radioactive-everyday-products-608655.
4. Todd Helmenstine, "List of Radioactive Elements and Their Most Stable Isotopes," ThoughtCo., January 27, 2019, https://www.thoughtco.com/list-of-radioactive-elements-608644.

5. Technuclear, "Radioactive Decay," accessed February 14, 2019, http://teachnuclear.ca/all-things-nuclear/radiation/radioactive-decay.
6. Ibid.
7. David Szondy, "Space Telescope Captures Storm-Generated Gamma-Ray Bursts," *New Atlas*, April 25, 2017, https://newatlas.com/gamma-rays-storms-nasa/49188/.
8. Brett F. Thornton and Shawn C. Burdette, "Neutron Stardust and the Elements of Earth," *Nature Chemistry* 11 (2019), https://www.nature.com/articles/s41557-018-0190-9.
9. Helmenstine.
10. Katrina Kramer, "Explainer: Superheavy Elements," *Chemistry World*, June 9, 2016, https://www.chemistryworld.com/news/explainer-superheavy-elements/1010345.article.
11. Helmenstine.
12. Ibid.

CHAPTER 5.
GAMMA RAYS IN SPACE

1. Jim Lucas, "What Are Gamma-Rays?" Live Science, November 29, 2018, https://www.livescience.com/50215-gamma-rays.html.
2. Goddard Space Flight Center, NASA, "What Is a Solar Flare?," accessed February 14, 2019, https://hesperia.gsfc.nasa.gov/sftheory/flare.htm.
3. Goddard Space Flight Center, NASA, "What Is a Solar Flare?"
4. Erika K. Carlson, "Blazars Explained," *Astronomy*, July 12, 2018, http://astronomy.com/news/2018/07/what-is-a-blazar.

5. Brandon Specktor, "Einstein and the Incredible Hulk Now Have Their Own Constellations (But You'll Never See Them)," Live Science, October 19, 2018, https://www.livescience.com/63879-fermi-gamma-ray-constellations-named.html.
6. Nasa Science, "Gamma Rays," accessed February 14, 2019, https://science.nasa.gov/ems/12_gammarays.
7. Ibid.
8. Christopher Crockett, "What Are Gamma Ray Bursts?," *Space*, October 14, 2013, https://earthsky.org/space/gamma-ray-bursts-are-the-most-powerful-explosions-in-the-universe.
9. Ibid.
10. Ernie Tretkoff, "Did Gamma Rays Cause Ordovician Mass Extinction?" APS News, July 2004, https://www.aps.org/publications/apsnews/200407/extinction.cfm.

CHAPTER 6.
GAMMA RAYS AT WORK

1. Lawrence Berkeley National Laboratory, "Industrial Applications," accessed February 14, 2019, https://www2.lbl.gov/abc/wallchart/chapters/13/6.html.
2. BGS, "Radiation Crosslinking," accessed February 14, 2019, https://en.bgs.eu/wp-content/uploads/2017/02/BGS_radiation_crosslinking_en-1.pdf.
3. Ibid.
4. Minerals North, "Gamma Ray Spectrometry," accessed February 14, 2019, http://mineralsnorth.ca/images/uploads/pdf/2017_Shives_Gamma_Ray_Spectrometry.pdf.

5. Center for Nuclear Science and Technology Information, "Art and Science Application," accessed February 14, 2019, http://nuclearconnect.org/know-nuclear/applications/art-and-science.
6. FDA.gov, "Food Irradiation: What You Need to Know," accessed February 14, 2019, https://www.fda.gov/Food/ResourcesForYou/Consumers/ucm261680.htm.
7. International Atomic Energy Agency, "Sterile Insect Technique," accessed February 14, 2019, https://www.iaea.org/topics/sterile-insect-technique.
8. *Encyclopaedia Britannica,* "Radiation," accessed February 14, 2019, https://www.britannica.com/science/radiation.
9. Maeve White, "Radiation in Agriculture," Stanford, March 16, 2017, http://large.stanford.edu/courses/2017/ph241/white-m2/.

CHAPTER 7.
GAMMA RAYS IN MEDICINE

1. Center for Nuclear Science and Technology Information, "Medical Applications," accessed February 14, 2019, http://nuclearconnect.org/know-nuclear/applications/medical-uses.
2. *Encyclopaedia Britannica*, "Positron Emission Tomography," accessed February 14, 2019, https://www.britannica.com/topic/positron-emission-tomography.
3. Steris, "Gamma Irradiation Processing," accessed February 14, 2019, https://www.steris-ast.com/services/gamma-irradiation/.

4. Mayo Clinic, "Brain Stereotactic Radiosurgery," accessed February 14, 2019, https://www.mayoclinic.org/tests-procedures/brain-stereotactic-radiosurgery/about/pac-20384679.
5. C. Randall Harrell, Valentin Djonov, Crissy Fellabaum, and Vladislav Volarevic, "Risks of Using Sterilization by Gamma Radiation: The Other Side of the Coin," *International Journal of Medical Sciences*, January 18 2018, https://www.ncbi.nlm.nih.gov/pmc/articles/PMC5820857/#B5.
6. Stanford Environmental Health & Safety, "Radiation Protection Guidance for Hospital Staff," June 29, 2017, https://ehs.stanford.edu/wp-content/uploads/sops/Hospital-Guidance-document.pdf.
7. Ibid.

GLOSSARY

ANTIMATTER Matter made up of antiparticles, which have the opposite charge and magnetic orientation of regular subatomic particles.

COSMIC RAY A high-speed particle or stream of particles that travels through space.

DECAY CHAIN A series of radioactive decays.

ISOMER Two or more compounds, radicals, or ions that have the same number of protons and electrons but differ in energy and manner of radioactive decay.

MECHANICAL WAVE A wave that requires a medium.

NEBULA A cloud of gas or dust in space.

NEUTRINO A subatomic particle with no charge and very small mass.

NOVA An exploding star that suddenly gets very bright.

PARTICLE ACCELERATOR A device that produces a beam of fast-moving charged particles.

PHOTOCATHODE A device that emits electrons when struck by light.

POLYMER A substance composed of very large molecules made up of chemical units called monomers.

RADIATION Energy emitted in the form of waves or particles.

SEMICONDUCTOR A substance whose conductivity is between that of a conductor and an insulator.

SUBATOMIC PARTICLES The fundamental pieces of matter.

SUPERNOVA A violently exploding star that increases its light output by many millions of times.

TRANSMUTE To change in form.

FURTHER READING

BOOKS

Cooper, Christopher. *The Basics of Nuclear Physics*. New York, NY: Rosen Publishing, 2015.

Dickmann, Nancy. *Energy from Nuclear Fission: Splitting the Atom*. St. Catharines, ON: Crabtree Publishing Company, 2016.

Hagler, Gina. *Discovering Quantum Mechanics*. New York, NY: Rosen Publishing, 2015.

Malam, John, and David West. *A Brief Illustrated History of Science*. North Mankato, MN: Capstone Press, 2017.

Stille, Darlene R. *Observatories in Space*. Chicago, IL: World Book, 2017.

WEBSITES

Know Nuclear

nuclearconnect.org/know-nuclear

Learn more about how nuclear energy is created and used.

Physics Central

physicscentral.com

Explore the world of physics.

Tour of the Electromagnetic Spectrum: Gamma Rays

science.nasa.gov/ems/12_gammarays

NASA describes where gamma rays come from and how they are detected.

INDEX